Producers

by Grace Hansen

BEGINNING SCIENCE: ECOLOGY

Abdo Kids Jumbo is an Imprint of Abdo Kids
abdobooks.com

abdobooks.com

Published by Abdo Kids, a division of ABDO, P.O. Box 398166, Minneapolis, Minnesota 55439.
Copyright © 2020 by Abdo Consulting Group, Inc. International copyrights reserved in all countries.
No part of this book may be reproduced in any form without written permission from the publisher.
Abdo Kids Jumbo™ is a trademark and logo of Abdo Kids.

Printed in the United States of America, North Mankato, Minnesota.

102019

012020

THIS BOOK CONTAINS
RECYCLED MATERIALS

Photo Credits: iStock, Science Source, Shutterstock

Production Contributors: Teddy Borth, Jennie Forsberg, Grace Hansen
Design Contributors: Dorothy Toth, Pakou Moua

Library of Congress Control Number: 2019941233
Publisher's Cataloging-in-Publication Data

Names: Hansen, Grace, author.

Title: Producers / by Grace Hansen

Description: Minneapolis, Minnesota : Abdo Kids, 2020 | Series: Beginning science: ecology |
    Includes online resources and index.

Identifiers: ISBN 9781532188978 (lib. bdg.) | ISBN 9781644942703 (pbk.) |
    ISBN 9781532189463 (ebook) | ISBN 9781098200442 (Read-to-Me ebook)

Subjects: LCSH: Plants, Edible--Juvenile literature. | Photosynthesis--Juvenile literature. | Plants--Juvenile
    literature. | Food webs (Ecology)--Juvenile literature. | Ecology--Juvenile literature.

Classification: DDC 577.16--dc23

Table of Contents

What Is a Producer?

All living things need energy to grow and survive. Humans and animals eat food to get energy. In the food chain, they are called consumers.

5

Plants are also living things.
But they do not have mouths
to eat with.

In the food chain, plants are called producers. This is because they make their own food. They do this through a process called **photosynthesis**.

sunlight

sugar and oxygen

carbon dioxide

water

9

Photosynthesis

Light from the sun is a form of energy. Plants capture the sun's light. They **convert** that energy into sugar.

Sugar is food for plants. They use it to grow and stay strong.

Producers support every single **ecosystem**. They do this by adding energy and oxygen to it.

Even plants that grow underwater are producers. They add oxygen to water. The plants are **shelter** for fish. They are food for other animals.

Producers in The Food Chain

All consumers get energy from producers. This can happen directly and indirectly. Many animals eat just plants for energy. Horses, koalas, and prairie dogs are plant eaters.

Some animals eat plants and other animals. And some just eat other animals. This is how energy flows through a food chain. But the chain always begins with the sun and a plant.

Let's Review!

- All living things need energy to survive.

- Plants are alive. They get energy by making their own food. They do this using sunlight, water, and carbon dioxide to make sugar.

- The process of plants making their own food is called **photosynthesis**.

- In the food chain, plants are called producers. This is because they produce their own food. And they are also food for other animals.

- Producers also make and give off oxygen. The oxygen goes into the air and water. Humans and animals breathe this oxygen.

Glossary

convert – to change into another form or state.

ecosystem – a community of living things, together with their environment.

photosynthesis – the process by which a green plant uses sunlight to change water and carbon dioxide into food for itself.

shelter – a place that gives protection against weather or danger.

Index

Abdo Kids
ONLINE
FREE! ONLINE MULTIMEDIA RESOURCES

Visit **abdokids.com**
to access crafts, games,
videos, and more!

Use Abdo Kids code
BPK8978
or scan this QR code!